South America

South America presents an overview of the geography of this continent and the countries that make up South America. The teaching and learning in this unit are based on the five themes of geography developed by the Association of American Geographers together with the National Council for Geographic Education.

The five themes of geography are described on pages 2 and 3. The themes are also identified on all student worksheets throughout the unit.

South America is divided into seven sections.

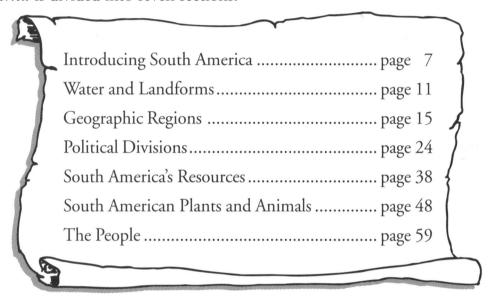

Each section includes:
* teacher resource pages explaining the activities in the section
* information pages for teachers and students
* reproducible resources
    maps
    note takers
    activity pages

Pages 4–6 provide suggestions on how to use this unit, including instructions for creating a geography center.

Congratulations on your purchase of some of the finest teaching materials in the world.

For information about other Evan-Moor products, call 1-800-777-4362 or FAX 1-800-777-4332

http://www.evan-moor.com

Author:	Jo Ellen Moore
Editor:	Jill Norris
Copy Editor	Cathy Harber
Desktop:	Keli Winters
Illustrator:	Cindy Davis
	Keli Winters
Cover Design:	Cheryl Puckett
Photography:	David Bridge, Mike Verbois, and Digital Stock

Entire contents copyright ©1999 by EVAN-MOOR CORP.
18 Lower Ragsdale Drive, Monterey, CA 93940-5746
Permission is hereby granted to the individual purchaser to reproduce student materials in this book for noncommercial individual or classroom use only. Permission is not granted for school-wide, or system-wide, reproduction of materials. Printed in U.S.A.

EMC 764

The Five Themes of Geography

Location

Position on the Earth's Surface

Location can be described in two ways. **Relative location** refers to the location of a place in relation to another place. **Absolute location** (exact location) is usually expressed in degrees of longitude and latitude.

> Uruguay is located south of the equator on the Atlantic Ocean.
> Brazil and Argentina form its other two borders.

> The absolute location of Cuzco, Peru, is 14°S latitude, 72°W longitude.

Place

Physical and Human Characteristics

Place is expressed in the characteristics that distinguish a location. It can be described in **physical characteristics** such as water and landforms, climate, etc., or in **human characteristics** such as languages spoken, religion, government, etc.

> While people in most of the countries of South America speak Spanish, Portuguese is the official language of Brazil.

Relationships within Places

Humans and the Environment

This theme includes studies of how people depend on the environment, how people adapt to and change the environment, and the impact of technology on the environment. Cities, roads, planted fields, and terraced hillsides are all examples of man's mark on a place. A place's mark on man is reflected in the kind of homes built, the clothing worn, the work done, and the foods eaten.

> Food, homes, and clothing for thousands of Amerindians come from the natural resources of the lush forests of the Amazon basin.

Movement

Human Interactions on the Earth

Movement describes and analyzes the changing patterns caused by human interactions on the Earth's surface. Everything moves. People migrate, goods are transported, and ideas are exchanged. Modern technology connects people worldwide through advanced forms of communication.

> Thousands of acres of the rainforest are destroyed each year as the expanding human population moves into the area.

Regions

How They Form and Change

Regions are a way to describe and compare places. A region is defined by its common characteristics and/or features. It might be a geographic region, an economic region, or a cultural region.

> Geographic region: The Atacama Desert is a high, arid region in northern Chile.
>
> Economic region: The coffee-raising areas of South America provide one of the country's most important exports.
>
> Cultural region: The Waorani are an Amerindian people living a traditional lifestyle in the Ecuadoran rainforest.

Using This Geography Unit

Good Teaching with *South America*

Use your everyday good teaching practices as you present material in this unit.

* Provide necessary background and assess student readiness:
    review necessary skills such as using latitude, longitude, and map scales
    model new activities
    preview available resources
* Define the task on the worksheet or the research project:
    explain expectations for the completed task
    discuss evaluation of the project
* Guide student research:
    provide adequate time for work
    provide appropriate resources
* Share completed projects and new learnings:
    correct misconceptions and misinformation
    discuss and analyze information

Doing Student Worksheets

Before assigning student worksheets, decide how to manage the resources that you have available. Consider the following scenarios for doing a page that requires almanac or atlas research:

* You have one classroom almanac or atlas.
    Make an overhead transparency of the page needed and work as a class to complete the activity, or reproduce the appropriate almanac page for individual students. (Be sure to check copyright notations before reproducing pages.)
* You have several almanacs or atlases.
    Students work in small groups with one resource per group, or rotate students through a center to complete the work.
* You have a class set of almanacs or atlases.
    Students work independently with their own resources.

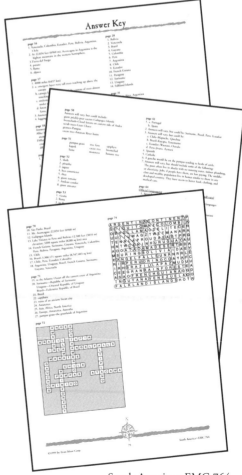

Checking Student Work

A partial answer key is provided on pages 77–79. Consider the following options for checking the pages:

* Collect the pages and check them yourself. Then have students make corrections.
* Have students work in pairs to check and correct information.
* Discuss and correct the pages as a class.

Creating a Geography Center

Students will use the center to locate information and to display their work.

Preparation

1. Post the unit map of South America on an accessible bulletin board.
2. Add a chart for listing facts about South America as they are learned.
3. Allow space for students to display newspaper and magazine articles on the continent, as well as samples of their completed projects.
4. Provide the following research resources:
   * world map
   * globe
   * atlas (one or more)
   * current almanac
   * computer programs and other electronic resources
   * fiction and nonfiction books (See bibliography on page 80.)
5. Provide copies of the search cards (pages 69–71), crossword puzzle (pages 72 and 73), and word search (page 74). Place these items in the center, along with paper and pencils.

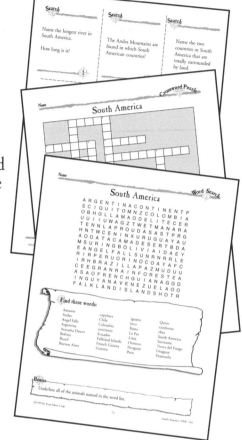

Additional Resources

At appropriate times during the unit, you will want to provide student access to these additional research resources:

* Filmstrips, videos, and laser discs
* Bookmarked sites on the World Wide Web (For suggestions, go to http://www.evan-moor.com and click on the Product Updates link on the home page.)

Making a Portfolio on South America

Provide a folder in which students save the work completed in this unit. Reproduce the following portfolio pages for each student:

* A Summary of Facts about South America, page 66
  Students will use this fact sheet to summarize basic information they have learned about South America. They will add to the sheet as they move through the unit.

* What's Inside This Portfolio?, page 67
  Students will record pages and projects that they add to the portfolio, the date of each addition, and why it was included.

* My Bibliography, page 68
  Students will record the books and other materials they use throughout their study of South America.

At the end of the unit have students create a cover illustration showing some aspect of South America.

Encourage students to refer to their portfolios often. Meet with them individually to discuss their learning. Use the completed portfolio as an assessment tool.

Using the Unit Map

Remove the full-color unit map from the center of this book and use it to help students do the following:

* locate and learn the names of landforms, water forms, and physical regions of South America
* practice finding relative locations using cardinal directions shown on the compass rose
* calculate distances between places using the scale

Introducing South America

Tour the Geography Center

Introduce the Geography Center to your class. Show the research materials and explain their uses. Ask students to locate the sections of atlases and almanacs containing material about South America.

Ecuadoran

Thinking about South America

Prepare a KWL chart in advance. Reproduce page 8 for each student. Give students a period of time (5–10 minutes) to list facts they already know about South America and questions about the continent they would like answered.

Know	Want to Know	Learned

Transfer their responses to the KWL chart. Post the chart in a place where you can add to it throughout your study of the continent.

Where Is South America?

Reproduce pages 9 and 10 for each student.

"Locating South America" helps students locate South America using relative location. Use the introductory paragraph to review the definition of relative location, and then have students complete the page.

"Name the Hemisphere" reviews the Earth's division into hemispheres. Students are asked to name the hemispheres in which South America is located. Using a globe to demonstrate the divisions, read the introduction together. Then have students complete the page.

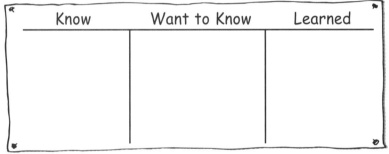

South America

What do you already know about the unique and fascinating continent of South America?

_____
_____
_____
_____
_____

_____
_____
_____
_____

If you could talk to someone from South America, what would you ask?

_____
_____
_____
_____
_____

_____
_____
_____
_____

Locating South America

Relative location tells where a place is located in relation to other places. Use the description of its relative location to help you find South America on the world map. Color in the continent on the map below and write South America on it.

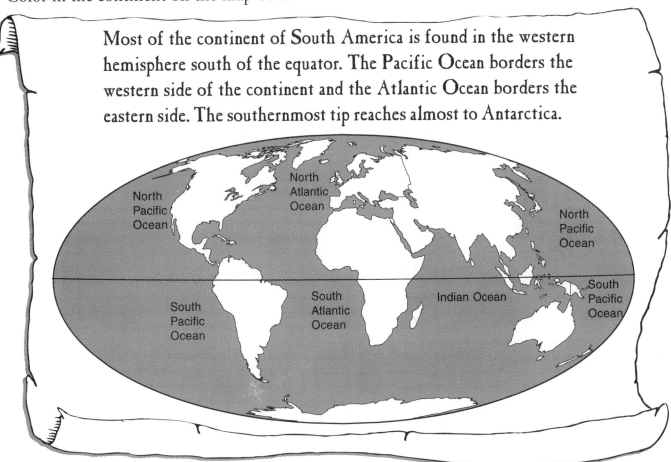

Most of the continent of South America is found in the western hemisphere south of the equator. The Pacific Ocean borders the western side of the continent and the Atlantic Ocean borders the eastern side. The southernmost tip reaches almost to Antarctica.

Look at a map of South America. Find these places and write their relative locations:

1. Chile _____

2. Amazon River _____

3. Falkland Islands _____

Bonus

Imagine you are describing the relative location of the state or territory in which you live to a student in South America. What would you say?

Name the Hemisphere

The globe can be divided in half two ways. Each half is called a **hemisphere**. When it is divided at the equator, the southern and northern hemispheres are created. When it is divided along the prime meridian and 180° longitude, the western and eastern hemispheres are created.

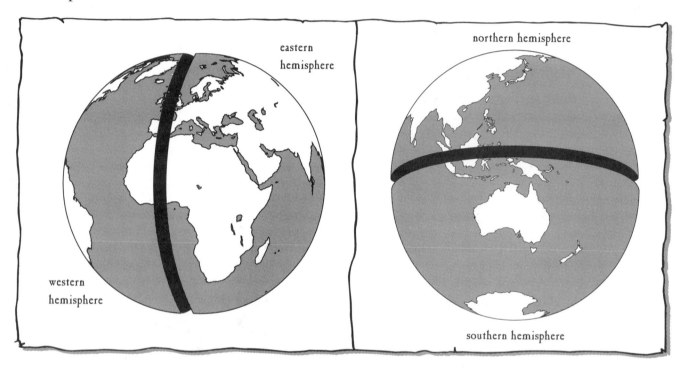

Look at South America on a globe. Then answer these questions:

1. In which hemispheres will you find South America? _____

2. Is more of South America found in the northern hemisphere or the southern hemisphere?

    _____

3. Which countries in South America are in the northern hemisphere?

    _____

Bonus

In which hemispheres do you live?

Water and Landforms

Collecting information by reading physical maps involves many skills. Pages 12–14 provide students with the opportunity to refine these skills as they learn about the water and landforms on the continent of South America.

Water Forms

Reproduce pages 12 and 13 for each student. Use the unit map to practice locating oceans, seas, lakes, and rivers on a map.

- Review how rivers and lakes are shown on a map.
- Discuss pitfalls students may face in finding the correct names (names written along the rivers, small type, several names close together).
- Have students locate at least one example of each type of water form on the unit map.
- Then have students locate and label the listed water forms on their individual physical maps.

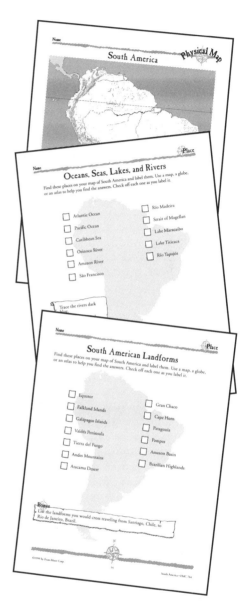

Landforms

Reproduce page 14 for each student. Have students use the same map used to complete page 13, or reproduce new copies of page 12 for this activity.

- Review the ways mountains, deserts, and other landforms are shown on a map (symbols, color variations, labels).
- Have students practice locating some of the mountains, deserts, and other landforms on the unit map of South America.
- Then have students locate and label the listed landforms on their individual physical maps.

South America

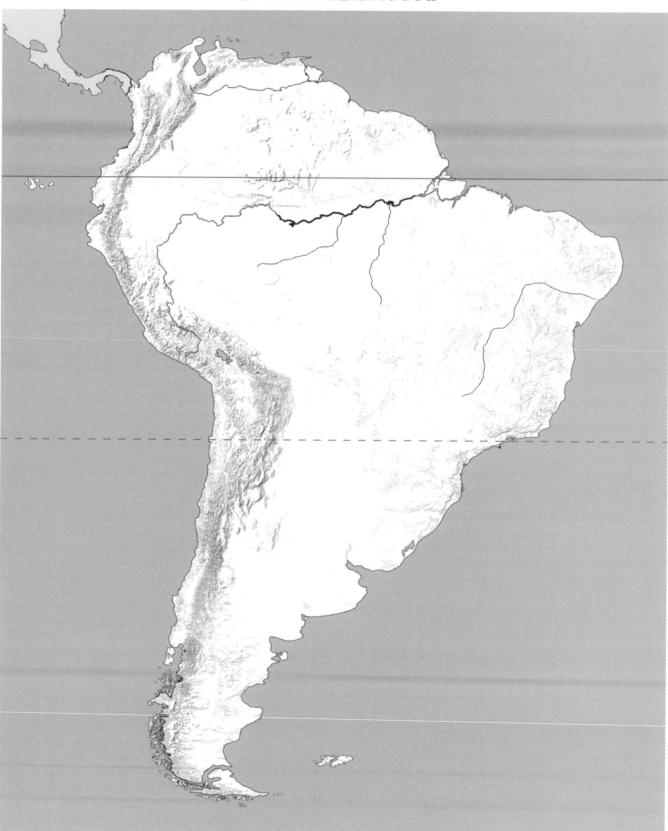

South America • EMC 764

Oceans, Seas, Lakes, and Rivers

Find these places on your map of South America and label them. Use a map, a globe, or an atlas to help you find the answers. Check off each one as you label it.

☐ Atlantic Ocean   ☐ Río Madeira

☐ Pacific Ocean   ☐ Strait of Magellan

☐ Caribbean Sea   ☐ Lake Maracaibo

☐ Orinoco River   ☐ Lake Titicaca

☐ Amazon River   ☐ Río Tapajós

☐ São Francisco River

Trace the rivers dark blue.

Color the lakes dark blue.

Color the seas and oceans light blue.

Bonus

Imagine you are traveling in a ship. Explain the route you would follow to get from Cape Horn to Manaus, Brazil, on the Amazon River.

South American Landforms

Find these places on your map of South America and label them. Use a map, a globe, or an atlas to help you find the answers. Check off each one as you label it.

☐ Equator	☐ Gran Chaco
☐ Falkland Islands	☐ Cape Horn
☐ Galápagos Islands	☐ Patagonia
☐ Valdés Peninsula	☐ Pampas
☐ Tierra del Fuego	☐ Amazon Basin
☐ Andes Mountains	☐ Brazilian Highlands
☐ Atacama Desert	

Bonus

List the landforms you would cross traveling from Santiago, Chile, to Rio de Janeiro, Brazil.

N

Geographic Regions

South America contains a huge mountain range, vast areas of rainforests, high plateaus, grasslands, and the driest desert on Earth. Each has distinct physical characteristics and climatic conditions. The material on pages 16–23 explores some of these regions.

Regions of South America

Reproduce pages 16–20 for each student and make an overhead transparency of page 21. As a class, discuss the material about each physical region while referring to the transparency. Share additional information from books and videos in your geography center. Then have students complete the pages, answering questions and coloring in the appropriate region on the physical map of South America (page 12).

Andes Mountains, page 16
Amazon Rainforest, page 17
Gran Chaco and Pampas, page 18
Atacama Desert, page 19
Altiplano, page 20

Comparing Regions

Reproduce page 22 for each student. Students are to fill in the chart to compare and contrast characteristics of two South American regions. They should recall the information they learned in the previous activity and do additional research using materials provided in the geography center.

Then have students choose one of the two regions, synthesize the information they have noted on the "Region Comparison Chart," and write a report about the region.

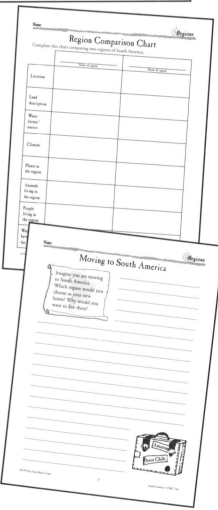

Moving to South America

As a summary activity on South America's regions, reproduce page 23 for each student. Have students write about the region they would choose to live in if they were immigrating to South America.

Andes Mountains

South America contains the longest chain of mountains in the world. The Andes Mountains run almost parallel with the western coast of the continent for about 4500 miles (7240 km). In some places the mountain range is more than 500 miles (805 km) wide. Many of the mountain peaks in this chain are very high. Some of the tallest peaks are volcanoes. To the west of the Andes is a narrow coastal strip of land.

Llama

Answer these questions:

1. Name the countries in which the Andes Mountains are found.

_____

2. Name the highest mountain in the Andes range. How tall is it?

_____

3. How far south on the continent do the Andes mountains reach?

_____

4. What commonly eaten vegetable was first domesticated in the Andes of Peru, then carried to the Old World by Spanish explorers?

_____

5. Name the animal used as a beast of burden in the Andes. _____

6. Name the Andean animal raised for its fine wool. _____

Color in the Andes Mountains on the physical map of South America.

Amazon Rainforest

The largest tropical rainforest in the world is located in the Amazon River Basin. It is a warm, rainy, and humid region of lush vegetation, with more species of plants and animals than in all the rest of the world's ecosystems combined.

The Amazon rainforest contains a variety of hardwood trees, palm trees, and fern trees, along with other vegetation such as vines and flowers. Because very little light from the sun can reach the ground, there is not much undergrowth.

Answer these questions:

1. How long is the Amazon River that runs through the Amazon River Basin?

_____

2. Name and describe the typical layers of a rainforest.

   a. _____

   b. _____

   c. _____

   d. _____

3. Name two examples of each type of animal living in the Amazon rainforest.

   insect   _____   _____

   bird   _____   _____

   mammal   _____   _____

   reptile   _____   _____

   amphibian   _____   _____

4. What human activities have caused destruction of the Amazon rainforest?

_____

_____

_____

Color in the area of the Amazon rainforest on the physical map of South America.

Name _____

Gran Chaco and Pampas

There are large plains and grasslands in many parts of South America.

The Gran Chaco is a low, flat plain in south central South America. It covers about 250,000 square miles (647,500 sq km) in Paraguay and parts of Bolivia and Argentina. It is a hot, dry region with areas of scrub forest, grasslands, and spiny brush forests. While it is usually dry, it sometimes floods during the summer rainy season. The vegetation varies in different areas. For example, along the waterways there are tall palm reeds, while inland there are scrub forests and grasslands, and to the west are spiny brush and arid places with no vegetation.

The Pampas is a vast region of plains in central Argentina reaching from the Atlantic coast to the Andes Mountains. While it is mostly grasslands with few trees, the eastern Pampas is one of the most fertile parts of the country. In the dryer areas to the west and south, cattle, horses, and sheep are raised.

Use class resources to help you answer these questions:

• How are the Gran Chaco and the Pampas alike? _____

_____

_____

_____

• How are the Gran Chaco and the Pampas different? _____

_____

_____

_____

Mark these areas on the physical map of South America:
1. Gran Chaco
2. Pampas

Name

Atacama Desert

The Atacama Desert is an arid region in northern Chile. The Atacama is located on a plateau between the Pacific Ocean and the Andes Mountains. It is about 100 miles (161 km) wide and about 600 miles (966 km) long. The desert is one of the driest places on Earth. Some parts of this desert have gone without rain for 400 years. Because of its altitude, the Atacama is a fairly cold desert.

There is sparse vegetation over most of the desert. Although there is little rainfall, there is some water. There are salt lakes where water accumulated long ago, small patches of snow, underground water, fog, and dew. These provide enough moisture for the few plants and animals that live in the Atacama. There are also a few oases that support small farming communities. Farmers in these communities raise cattle and grow crops using irrigation.

Use the information on this page to locate the Atacama Desert on this map. Color in the area of the desert and label it.

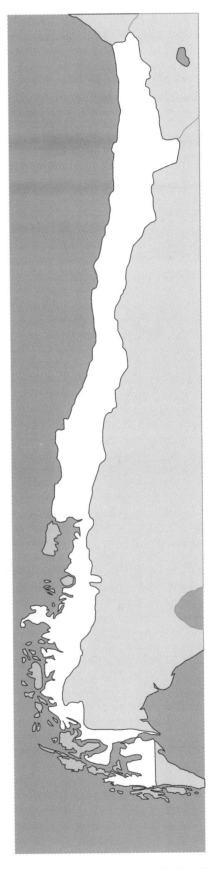

Bonus

Compare the Atacama Desert with a desert in your home country.

Altiplano

The Altiplano or "high plain" extends from southwestern Bolivia into southern Peru. The plain is a plateau at an altitude of about 12,000 feet (3650 m) and contains some of the most rugged terrain in Bolivia. The land varies from high, open grasslands to mountain peaks with glaciers. It has a cold, dry, and windy climate. In spite of this, many people live there farming and raising livestock. Along the Bolivian-Peruvian border is one of the world's highest navigable lakes—Lake Titicaca.

There are many interesting animals living on the Altiplano. Use class resources to find out about the following animals. Describe each animal and write an interesting fact about it.

1. James flamingo _____

_____

2. Andean condor _____

_____

3. Vicuña _____

_____

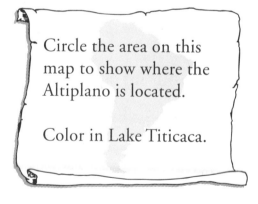

Circle the area on this map to show where the Altiplano is located.

Color in Lake Titicaca.

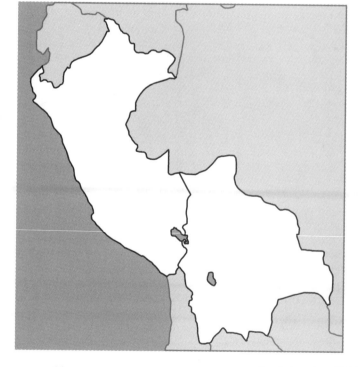

Geographic Regions of South America

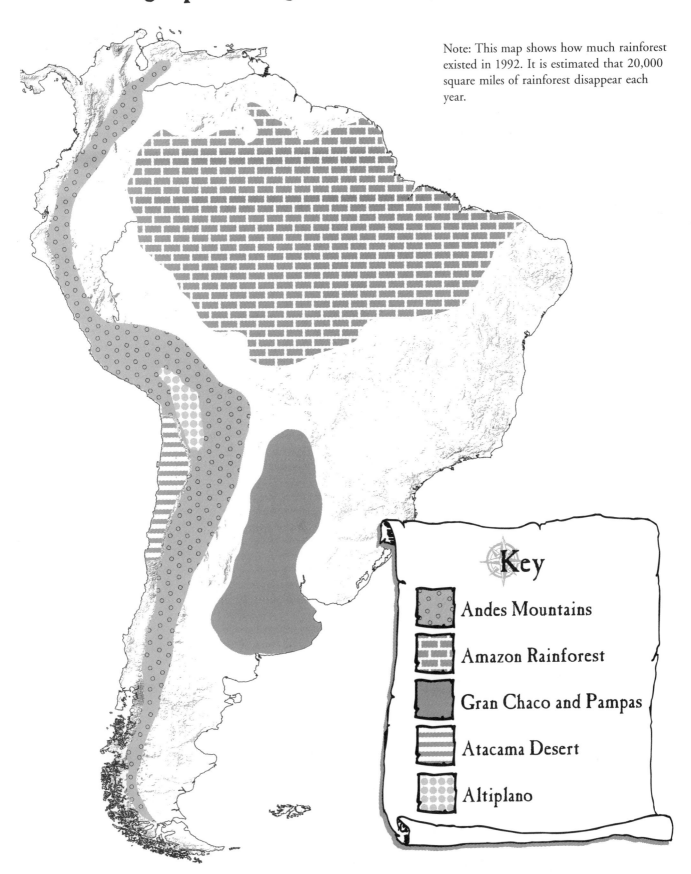

Note: This map shows how much rainforest existed in 1992. It is estimated that 20,000 square miles of rainforest disappear each year.

Key

- Andes Mountains
- Amazon Rainforest
- Gran Chaco and Pampas
- Atacama Desert
- Altiplano

Region Comparison Chart

Complete this chart comparing two regions of South America.

	Name of region	Name of region
Location		
Land description		
Water forms/ source		
Climate		
Plants in the region		
Animals living in the region		
People living in the region		
Ways people have changed the region		

Moving to South America

Imagine you are moving to South America. Which region would you choose as your new home? Why would you want to live there?

Political Divisions

A political map shows boundaries between countries or between states and territories. In this section students will use political maps to learn the countries of South America and their capital cities, to calculate distance and direction, and to locate places using longitude and latitude.

Countries of South America

Reproduce pages 26–28 for each student. Have students use map resources to do the following:

* list the countries in South America
* label the countries on the political map
* find the country is which each listed capital city is located
* write the capital cities on the political map

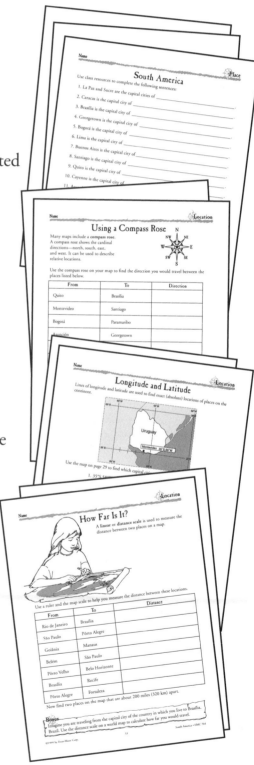

Using a Compass Rose

Reproduce pages 29 and 30 for each student. Use the compass rose on the unit map to review how to determine location using cardinal directions. Then have students complete the activity independently.

Longitude and Latitude

Reproduce pages 29 and 31 for each student. Use a map to review how to use lines of longitude and latitude to determine exact locations. Then have students complete the activity independently.

How Far Is It?

Reproduce pages 32 and 33 for each student. Use the unit map to review how to use a map scale to figure distances. Then have students use a ruler and the map scale to determine the distance between various cities in Brazil.

Land Area and Population Comparison

Reproduce pages 34 and 35 for each student. Send students to the geography center to find the size of each South American country by land area and population. Students use the information to rank the countries by size from largest to smallest, and then answer questions using the information.

Name the Countries

Reproduce page 36 for each student. Review the term "relative location." Have students use maps and atlases to name the countries described. They are then to write the relative location of several South American countries.

Country Fact Sheet

Reproduce page 37 for each student. Explain that you would like students to help create a file of fact sheets for the countries in South America. Have each student select a different country to research. Allow time for students to share what they discover about their countries. Keep the completed sheets in a binder in the geography center.

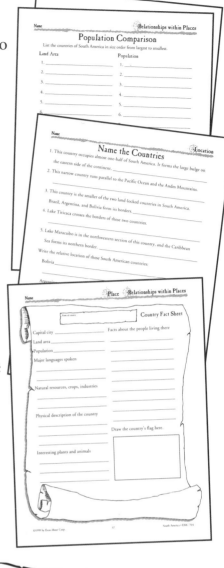

Ecuadoran

Scarlet Macaw

South America

Countries of South America

Use a map, an atlas, or a globe to help you with this activity. List the names of the countries of South America. Then find the location of each country and label it on your map of South America.

Independent Countries:

1. _____

2. _____

3. _____

4. _____

5. _____

6. _____

7. _____

8. _____

9. _____

10. _____

11. _____

12. _____

French Dependency: _____

United Kingdom Dependency: _____

South America

Use class resources to complete the following sentences:

1. La Paz and Sucre are the capital cities of _____.

2. Caracas is the capital city of _____.

3. Brasília is the capital city of _____.

4. Georgetown is the capital city of _____.

5. Bogotá is the capital city of _____.

6. Lima is the capital city of _____.

7. Buenos Aires is the capital city of _____.

8. Santiago is the capital city of _____.

9. Quito is the capital city of _____.

10. Cayenne is the capital city of _____.

11. Asunción is the capital city of _____.

12. Paramaribo is the capital city of _____.

13. Montevideo is the capital city of _____.

14. Stanley is the capital city of _____.

Write the name of each capital city in the correct place on your political map.

Bonus

Name the capital city of your country and the capital city of your state, province, or territory.

South America

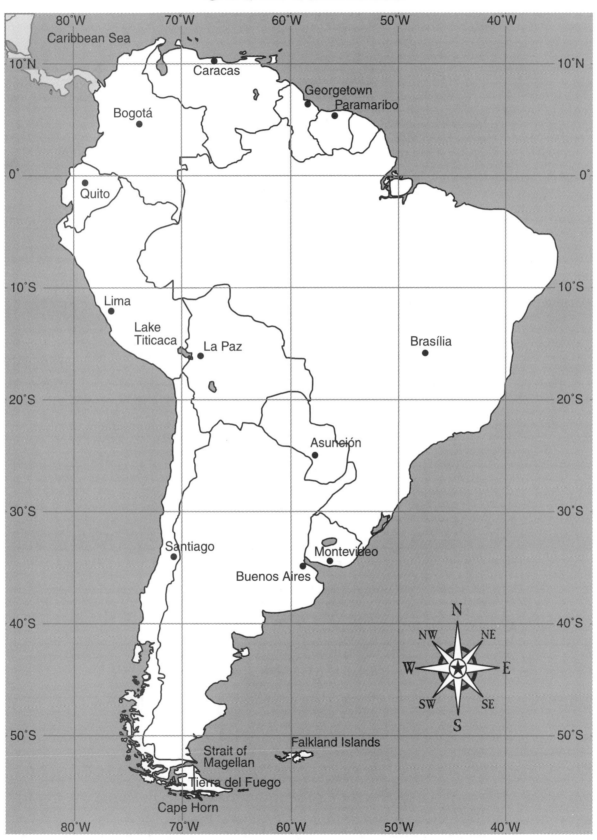

Caribbean Sea

80°W 70°W 60°W 50°W 40°W

10°N ● Caracas 10°N

Bogotá Georgetown
Paramaribo

0° 0°

● Quito

10°S 10°S

Lima ●
Lake
Titicaca La Paz ● Brasília ●

20°S 20°S

Asunción ●

30°S 30°S

Santiago ● Montevideo ●

Buenos Aires

N
NW NE
40°S W ★ E 40°S
SW SE
S

Falkland Islands
50°S 50°S

Strait of
Magellan

Tierra del Fuego

Cape Horn

80°W 70°W 60°W 50°W 40°W

Using a Compass Rose

Many maps include a **compass rose**.
A compass rose shows the cardinal
directions—north, south, east,
and west. It can be used to describe
relative locations.

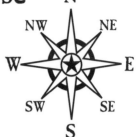

Use the compass rose on your map to find the direction you would travel between the
places listed below.

From	To	Direction
Quito	Brasília	
Montevideo	Santiago	
Bogotá	Paramaribo	
Asunción	Georgetown	
Santiago	Falkland Islands	
Caracas	Cape Horn	
Georgetown	Lake Titicaca	
Tierra del Fuego	Santiago	
Caribbean Sea	Strait of Magellan	
Sucre	La Paz	

Bonus

Use the cardinal directions to explain how to get from the capital city of the country in
which you live to the capital city of Ecuador.

Longitude and Latitude

Lines of longitude and latitude are used to find exact (absolute) locations of places on the continent.

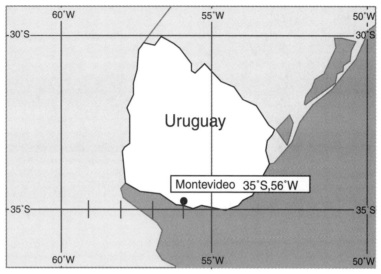

Use the map on page 29 to find which capital cities are located at these points:

1. 35°S,58°W   _____

2. 16°S,68°W   _____

3. 16°S,48°W   _____

4. 4°N,74°W   _____

5. 34°S,71°W   _____

6. 0°, 78°W   _____

7. 7°N,58°W   _____

8. 25°S,57°W   _____

9. 12°S,77°W   _____

10. 11°N,67°W   _____

Bonus

Write the location of the capital city of the country in which you live using longitude and latitude.

Brazil

Manaus

Belém

Fortaleza

Recife

Pôrto Velho

Brasília

Goiânia

Belo Horizonte

Rio de Janeiro

São Paulo

Pôrto Alegre

Key

Brazil

⊕ National Capital

● City

— International Boundary

400 Kilometers

0

400 Miles

·Location

How Far Is It?

A **linear** or **distance scale** is used to measure the distance between two places on a map.

Use a ruler and the map scale to help you measure the distance between these locations.

From	To	Distance
Rio de Janeiro	Brasília	
São Paulo	Pôrto Alegre	
Goiânia	Manaus	
Belém	São Paulo	
Pôrto Velho	Belo Horizonte	
Brasília	Recife	
Pôrto Alegre	Fortaleza	

Now find two places on the map that are about 200 miles (320 km) apart.

Bonus

Imagine you are traveling from the capital city of the country in which you live to Brasília, Brazil. Use the distance scale on a world map to calculate how far you would travel.

Name

Land Area and Population Chart

Use a current almanac to fill in the land area and population of each country in South America.

Country	Land Area	Population
Colombia		
Ecuador		
Peru		
Venezuela		
Suriname		
Guyana		
French Guiana		
Bolivia		
Paraguay		
Brazil		
Chile		
Uruguay		
Argentina		
Falkland Islands		

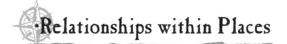

Population Comparison

List the countries of South America in size order from largest to smallest.

Land Area

1. _____
2. _____
3. _____
4. _____
5. _____
6. _____
7. _____
8. _____
9. _____
10. _____
11. _____
12. _____
13. _____
14. _____

Population

1. _____
2. _____
3. _____
4. _____
5. _____
6. _____
7. _____
8. _____
9. _____
10. _____
11. _____
12. _____
13. _____
14. _____

Which countries are in the same position on both lists?

_____

_____

Bonus

Give the land area and the population of the country in which you live.

Name the Countries

1. This country occupies almost one-half of South America. It forms the large bulge on

   the eastern side of the continent. _____

2. This narrow country runs parallel to the Pacific Ocean and the Andes Mountains.

   _____

3. This country is the smaller of the two land-locked countries in South America.

   Brazil, Argentina, and Bolivia form its borders. _____

4. Lake Titicaca crosses the borders of these two countries.

   _____

5. Lake Maracaibo is in the northwestern section of this country, and the Caribbean

   Sea forms its northern border. _____

Write the relative location of these South American countries:

   Bolivia_____

   _____

   Argentina _____

   _____

   Suriname_____

   _____

   Ecuador _____

   _____

Bonus

Imagine you are talking about your country to someone from Brazil. Describe its relative location.

Name of country

Country Fact Sheet

Capital city _____

Land area _____

Population _____

Major languages spoken

_____

_____

Natural resources, crops, industries

_____

_____

Physical description of the country

_____

_____

_____

Interesting plants and animals

_____

_____

_____

Facts about the people living there

_____

_____

_____

_____

_____

_____

_____

_____

_____

_____

Draw the country's flag here.

South America's Resources

The activities in this section introduce students to the natural and man-made resources of South America.

Resources

Prepare for this lesson by enlarging the political map on page 26, using an overhead projector and a sheet of butcher paper. Post the map on a bulletin board.

Marketplace

Reproduce page 40 for each student. Assign one country to each student. Have them use atlases, almanacs, and books to locate information about the natural resources, crops and livestock, and manufactured goods of the countries they have been assigned. Students are to record the information gathered on their activity pages.

Create a "key" on the map using symbols agreed upon by the students. Then have students place symbols for the items on their lists in the appropriate locations on the large map of South America.

Foods from the New World

Prepare for this lesson by gathering foods that originally came from South America (cocoa or chocolate, sweet and hot peppers, etc.). Reproduce pages 41 and 42 for each student.

Introduce the terms "New World" and "Old World" as they relate to the eastern and western hemispheres. Have students locate these areas on a world map. As a class, share the information about food plants that have moved from the New World to the Old World. Show students examples of the foods as you discuss them. Then have students use class resources to find the answers to the questions on the page.

You can extend the lesson in several ways:
* create a class graph showing who has eaten each type of food discussed in the lesson.
* have students do research to find out what foods are being imported from South America today. Send students on a food search at a local supermarket to find some of these foods. (Include packaged foods as well as fresh foods.)

Imports–Exports

Make an overhead transparency of page 43 and also reproduce a copy for each student. Discuss the terms "import" and "export" with students. As a class, write a definition of each term. Show the graph transparency as you compare the amount of money each country shown has spent on imports and made on exports (data is from 1996 figures). Ask students to think about the impact on a country if imports are always greater than exports. Have students use the information on the graph to answer the questions.

Deforestation

Reproduce pages 44 and 45 for each student. As a class, read and discuss the information, and then send students to class resources to learn more about the causes of this problem and what is being done to slow deforestation in the rainforests of South America. Have students record what they find on their note takers. Students should share what they discover, and then synthesize the information gathered to write a report.

Visit South America

Come to South America

Tourism is a big industry in many parts of South America. Visit a travel agency to get samples of brochures and posters about South American trips. After sharing these materials, have students develop one of the following:

* a brochure of things to do on a South American vacation
* a travel poster about one special place or site in South America
* a list of ways to be a considerate tourist
* a video advertisement encouraging people to come to South America

Responsible Tourism

Reproduce page 46 for each student. As a class, read and discuss the information. Send students to class resources to learn more about ecotourism in South America. Then have them answer the questions.

A South American Vacation

Reproduce page 47 for each student. Students are to select a place to visit in South America, explain their choice, and write a postcard they might send while on the imagined visit.

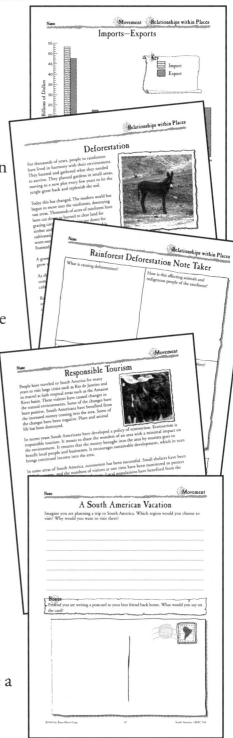

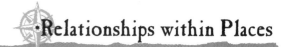

Resources of _____
country's name

Use atlases, maps, and other resources to develop a list of natural resources, crops and livestock, and manufactured goods found in this South American country.

Natural Resources	Crops and Livestock	Manufactured Goods

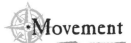

Foods from the New World

Potatoes are one of the world's main food crops. But until the discovery of the New World by European explorers, only people in parts of what is now South America had ever eaten them.

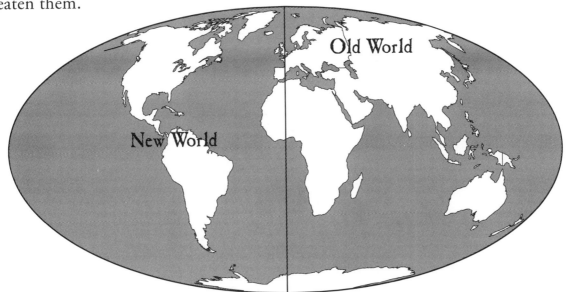

Potatoes are native plants of the Peruvian-Bolivian Andes. They have been cultivated for over 1800 years. When Spaniards arrived in Peru in the early 1500s, they were introduced to potatoes. Potatoes were sent back to Spain. From there potatoes gradually spread across Italy, France, and Great Britain. Eventually, potatoes were taken to the colonies in North America. Potatoes had traveled from the New World to the Old World and back again.

Potatoes are not the only food that came from South America. Before the European exploration of South America, people in the Old World had not eaten any of these foods:

peanuts   peppers   cacao (chocolate)   potatoes   squash   pineapple

Today these foods are grown in many parts of the world.

Bonus

Go to a supermarket and look for the foods on this list. See if there are any for sale that have been imported from South America.

My Personal Experiences
with Foods from the New World

Answer these questions:

1. What part of the peanut plant do you eat? _____

2. What part of the potato plant do you eat? _____

3. What is the mildest pepper? _____

4. What is the hottest pepper? _____

5. What plant is used to make chocolate? _____

6. Circle the foods you have eaten.

| peanuts | peppers | cacao (chocolate) | potatoes | squash | pineapple |

7. Number the foods in the order from the one you like best (#1) to the one you like least (#6).

_____   _____

_____   _____

_____   _____

8. Have you ever eaten a really hot pepper? If so, what did it feel like in your mouth?

_____

_____

Bonus

Imagine you were a sailor returning from the New World with a "new food." Describe the food in a way that would encourage others to try it.

Imports—Exports

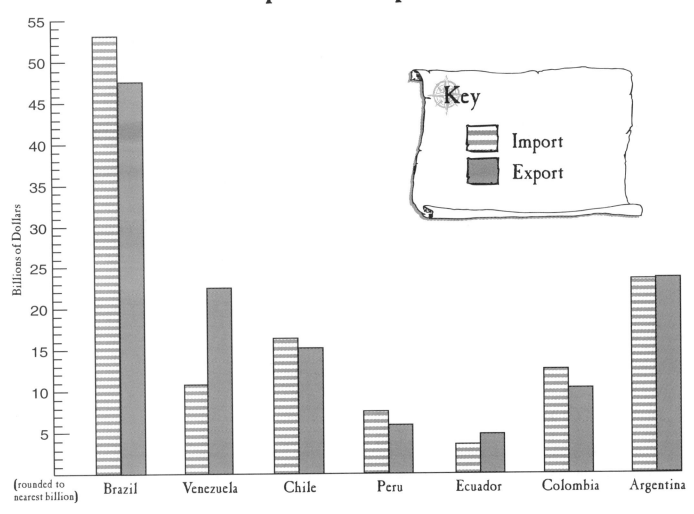

Key

Import
Export

(rounded to nearest billion)   Brazil   Venezuela   Chile   Peru   Ecuador   Colombia   Argentina

Read the graph to help you answer these questions:

1. Which countries import more than they export? _____

_____

2. Which countries import and export almost the same amount? _____

_____

3. Which countries export more than they import? _____

_____

Bonus

Find the export and import figures for the country in which you live.

Deforestation

For thousands of years, people in rainforests have lived in harmony with their environment. They hunted and gathered what they needed to survive. They planted gardens in small areas, moving to a new plot every few years to let the jungle grow back and replenish the soil.

Burro

Today this has changed. The modern world has begun to move into the rainforests, destroying vast areas. Thousands of acres of rainforest have been cut down or burned to clear land for grazing cattle. Trees are being cut down for timber and to clear room for roads. Land is cultivated and then abandoned when the worn-out soil will no longer produce crops. Forested areas have been cleared for the mining of minerals.

A growing population only adds to the problem. Towns and cities have begun to grow up along the edges of the rainforest. More land is cleared as more people arrive.

As the rainforest is cut down, all indigenous life is affected. Plants and animals have become extinct. Indigenous people have been moved from their traditional homes, and their cultures have been affected by interactions with the outside world.

Rainforests are vital to the well-being of our planet because of their effect on rainfall, their oxygen production, and the way they combat global warming. People all over the world have become aware of what is happening in rainforests today. Many organizations are working to find ways to stop further destruction of the rainforests.

Many countries in South America are trying to find ways to use the bounty of the rainforests in nondestructive ways. One way is to use sustainable products (products that can be harvested without killing the plant) such as fruits, nuts, latex, palm oils, cacao, and medicinal drugs. Another way is to develop better ways of logging and farming.

Rainforest Deforestation Note Taker

What is causing deforestation?	How is this affecting animals and indigenous people of the rainforest?
How is this affecting the rest of the world?	What is being done about the problem?

Bonus

Describe something you can do to help save the rainforests.

Responsible Tourism

People have traveled to South America for many years to visit large cities such as Rio de Janeiro and to marvel at lush tropical areas such as the Amazon River basin. These visitors have caused changes in the natural environments. Some of the changes have been positive. South Americans have benefited from the increased money coming into the area. Some of the changes have been negative. Plant and animal life has been destroyed.

In recent years South Americans have developed a policy of ecotourism. Ecotourism is responsible tourism. It means to share the wonders of an area with a minimal impact on the environment. It ensures that the money brought into the area by tourists goes to benefit local people and businesses. It encourages sustainable development, which in turn brings continued income into the area.

In some areas of South America, ecotourism has been successful. Small shelters have been built for guests, and the numbers of visitors at one time have been monitored to protect the plant, animal, and human populations. Local populations have benefited from the increased money coming into the area.

In other areas of South America, ecotourism has failed. Luxury hotels, shopping centers, and golf courses have been built for tourists. Advertising encourages large numbers of tourists to visit at any time. The natural habitats of animals have been destroyed by this development, while the money brought in by these tourists does not necessarily reach the local population.

Answer these questions:

1. What implications does increased tourism have for the South American continent?

   _____

   _____

2. Imagine that you want to explore one of South America's environments. How can you be a thoughtful ecotourist?

   _____

   _____

Name

A South American Vacation

Imagine you are planning a trip to South America. Which region would you choose to visit? Why would you want to visit there?

_____

_____

_____

_____

_____

_____

Bonus

Pretend you are writing a postcard to your best friend back home. What would you say on the card?

South American Plants and Animals

With environments as diverse as high mountains; humid rainforests; and hot, dry deserts, South America has an incredible variety of plant and animal life.

Plants of South America

Reproduce pages 50 and 51 for each student. Share appropriate sections of books and videos about South America. Divide students into small groups. Have them work together using class resources to create lists of plants found in the various environments of South America. Students should describe the habitat in which each plant is found. Provide time for students to share what they discover. Then have students identify the plants pictured on page 51.

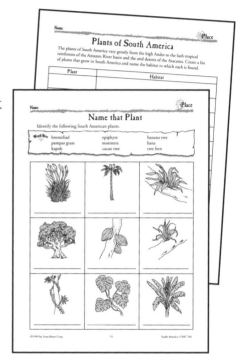

South American Animals

Introductory Activities

Begin by challenging students to name the South American animals they know. (You may need to name an area such as the rainforest or the Andes to get them started.) List the following animal names on a chart and write a descriptive phrase for each one.

> llama–a four-footed animal used to carry goods
> howler monkey–a large, noisy monkey living in the rainforest
> sloth–a slow-moving mammal with long hair
> rhea–a tall, flightless bird similar to an ostrich

Share books or show a video about South American animals. Discuss the information learned from these sources, and add new animal names and descriptive phrases to the chart.

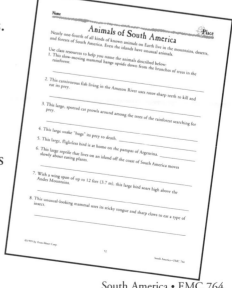

Animals of South America

Reproduce page 52 for each student. Have students use class resources to find the answers to the animal riddles.

South America • EMC 764

South American Camels

Reproduce page 53 for each student. As a class, read and discuss the information at the top of the page. Have students use class resources including the World Wide Web to find additional information. Then have students complete the page independently.

Note: You will want to rotate groups through the center if your resources are limited.

Amazing Animals of South America

Reproduce pages 54 and 55 for each student. Send students to class resources to locate two interesting facts about each animal shown.

Animals of the Galápagos Islands

Reproduce pages 56 and 57 for each student. As a class, read and discuss the plant and animal life found on the Galápagos Islands. Then have students complete the activity.

South American Animal Report

Provide each student with a copy of the note taker on page 58. Have students choose one interesting animal from South America and use class resources to complete their note takers. Then have students share the information in oral or written reports.

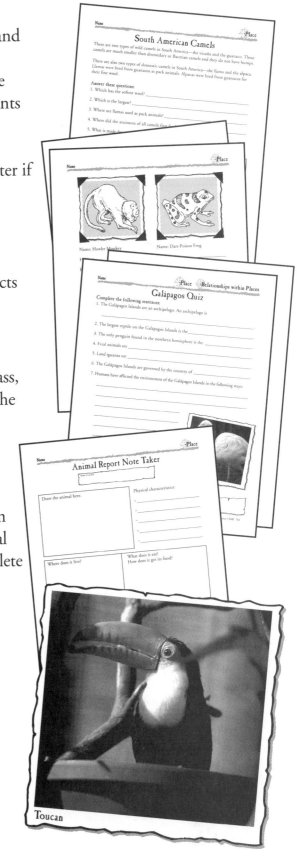

Toucan

Plants of South America

The plants of South America vary greatly from the high Andes to the lush tropical rainforests of the Amazon River basin and the arid deserts of the Atacama. Create a list of plants that grow in South America and name the habitat in which each is found.

Plant	Habitat

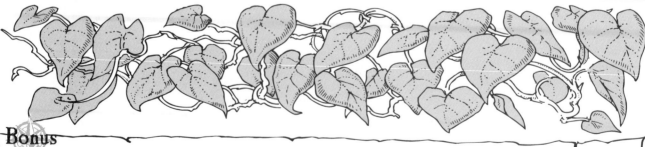

Bonus

Walk around your neighborhood to see how many plants growing there you can name.

Name that Plant

Identify the following South American plants.

Word Box   bromeliad   epiphyte   banana tree
            pampas grass   monstera   liana
            kapok tree   cacao tree   tree fern

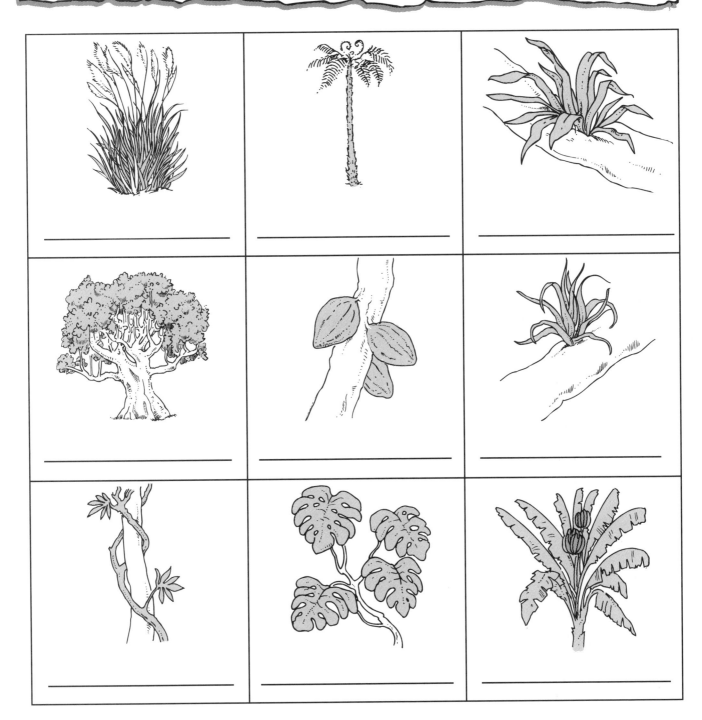

Animals of South America

Nearly one-fourth of all kinds of known animals on Earth live in the mountains, deserts, and forests of South America. Even the islands have unusual animals.

Use class resources to help you name the animals described below:

1. This slow-moving mammal hangs upside down from the branches of trees in the rainforest.

   _____

2. This carnivorous fish living in the Amazon River uses razor-sharp teeth to kill and eat its prey.

   _____

3. This large, spotted cat prowls around among the trees of the rainforest searching for prey.

   _____

4. This large snake "hugs" its prey to death. _____

5. This large, flightless bird is at home on the pampas of Argentina. _____

6. This large reptile that lives on an island off the coast of South America moves slowly about eating plants.

   _____

7. With a wing span of up to 12 feet (3.7 m), this large bird soars high above the Andes Mountains.

   _____

8. This unusual-looking mammal uses its sticky tongue and sharp claws to eat a type of insect.

   _____

South American Camels

There are two types of wild camels in South America—the vicuña and the guanaco. These camels are much smaller than dromedary or Bactrian camels and they do not have humps.

There are also two types of domestic camels in South America—the llama and the alpaca. Llamas were bred from guanacos as pack animals. Alpacas were bred from guanacos for their fine wool.

Answer these questions:

1. Which has the softest wool? _____

2. Which is the largest? _____

3. Where are llamas used as pack animals? _____

4. Where did the ancestors of all camels first live? _____

5. What is made from the wool of alpacas? _____

Name these South American camels.

Clue: I live in the wild in the foothills of the Andes Mountains.	Clue: My wool is made into cloth.
Clue: I am the largest South American camel.	Clue: I have the softest wool.

Amazing Animals of South America

Name: Capybara

Habitat: _____

Facts: _____

_____

Name: Piranha

Habitat: _____

Facts: _____

_____

Name: Blue-footed Booby

Habitat: _____

Facts: _____

_____

Name: Pampas Cat

Habitat: _____

Facts: _____

_____

Name: Howler Monkey

Habitat: _____

Facts: _____

_____

Name: Dart-Poison Frog

Habitat: _____

Facts: _____

_____

Name: Harpy Eagle

Habitat: _____

Facts: _____

_____

Name: Marine Iguana

Habitat: _____

Facts: _____

_____

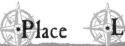

Animals of the Galápagos Islands

Giant Tortoise

The Galápagos Islands are about 621 miles (1000 km) off the western equatorial coast of South America. The islands are home to many unique animals, some of which are found no where else in the world. One of the most well-known inhabitants of this archipelago is the giant tortoise.

There are also many kinds of birds on the islands. For example, there is a woodpecker that uses tools, a type of hawk found no where else in the world, a flightless cormorant, and the only penguin that lives in the northern hemisphere.

Small rice rats and two kinds of bats are the only land mammals native to the islands. Other mammals such as goats, dogs, and cats were brought to the islands by early settlers and have become wild (feral). This has created a problem for the native plants and animals. The feral animals eat plants used by native tortoises and iguanas, prey on the chicks of sea birds, and eat the eggs of other animals.

Although there are few land mammals, many mammals are found in the sea around the islands. There are sea lions and fur seals along the coasts and dolphins, whales, and sea turtles farther offshore.

People have settled on some of the islands. These settlements have brought changes to the natural environment. Tourism, fishing, agriculture, and harvesting of timber have all had an impact. The major challenge facing the islands today is developing ways to meet the needs of an expanding population without destroying the existing environment.

Galápagos Quiz

Complete the following sentences:

1. The Galápagos Islands are an archipelago. An archipelago is

_____.

2. The largest reptile on the Galápagos Islands is the _____.

3. The only penguin found in the northern hemisphere is the _____.

4. Feral animals are _____.

5. Land iguanas eat _____.

6. The Galápagos Islands are governed by the country of _____.

7. Humans have affected the environment of the Galápagos Islands in the following ways:

_____

_____

_____

_____

_____

_____

_____

_____

_____

Flamingo

Bonus

Are there any feral animals in the community in which you live?

Animal Report Note Taker

Name of animal

Draw the animal here.	Physical characteristics:
	• _____
	• _____
	• _____
	• _____
	• _____

Where does it live?	What does it eat? How does it get its food?
How does it protect itself?	Describe its life cycle.

The _____ is/is not endangered.

The People

South America's People

Introduction

Invite speakers from South America or of South American descent to speak to the class. Prepare students for your speakers by planning questions to ask. Appoint several students to record questions asked and answers received. Follow up the visit by writing thank-you letters.

The People of South America

Reproduce pages 60 and 61 for each student. Define "ethnic groups" (having to do with a group of people who have the same race, nationality, or culture) and "cultural groups" (sharing the same customs, arts, language, etc.). As a class, read and discuss the information page. Students then use this information and class resources to complete the activity independently.

The Incas

Reproduce page 62 for each student. As a class, search for and share information about the ancient Incas. Students record notes on their note takers, and then synthesize what they learn into a written report. Encourage students to illustrate their reports with maps and pictures about the Incas. (This is an excellent opportunity to incorporate technology by preparing multimedia presentations.)

Indigenous Cultures

Reproduce page 63 for each student. Read and discuss the information at the top of the page together. Help students understand the concept of "culture." Divide students into small groups. Have each group select one indigenous culture to learn about. Students record information on their note takers, and then synthesize the information into oral or written reports. Provide time for students to share what they've learned with the rest of the class.

The Languages of South America

Reproduce page 64 for each student. Have them find the official language of each country and dependency in South America, and then answer questions to complete the page.

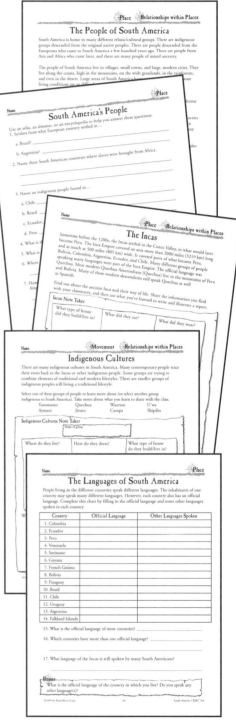

The People of South America

South America is home to many different ethnic/cultural groups. There are indigenous groups descended from the original native peoples. There are people descended from the Europeans who came to South America a few hundred years ago. There are people from Asia and Africa who came later, and there are many people of mixed ancestry.

The people of South America live in villages, small towns, and large, modern cities. They live along the coasts, high in the mountains, on the wide grasslands, in the rainforests, and even in the desert. Large areas of South America have sparse populations because living conditions are so difficult.

The culture of each country is a mixture of parts from the indigenous population, European settlers, and in some countries, African slaves. Customs, music, and art forms reflect aspects of each group of people.

As transportation and communication have become more widespread and new industries have developed, the way of life of many people is changing. Indigenous people are in contact with the outside world, bringing changes to their traditional way of life. Many people are immigrating from small villages to large cities searching for better jobs.

The people of South America face many challenges as their world changes, but they also face great opportunities.

Village Children

City People

South America's People

Use an atlas, an almanac, or an encyclopedia to help you answer these questions:

1. Settlers from what European country settled in...

   a. Brazil? _____

   b. Argentina? _____

2. Name three South American countries where slaves were brought from Africa.

   _____

   _____

3. Name an indigenous people found in...

   a. Chile _____

   b. Brazil _____

   c. Ecuador _____

   d. Peru _____

4. What is the most common language spoken in South America? _____

5. What is the most common religion in South America? _____

6. Where would you find a gaucho? What would he be doing? _____

   _____

7. How is life different for a poor person than for a middle-class or rich person in a South American city?

   _____

   _____

   _____

   _____

The Incas

Sometime before the 1200s, the Incas settled in the Cuzco Valley in what would later become Peru. The Inca Empire covered an area more than 2000 miles (3219 km) long and as much as 500 miles (805 km) wide. It covered parts of what became Peru, Bolivia, Colombia, Argentina, Ecuador, and Chile. Many different groups of people speaking many languages were part of the Inca Empire. The official language was Quechua. Most modern Quechua Amerindians (Quechua) live in the mountains of Peru and Bolivia. Many of these modern descendents still speak Quechua as well as Spanish.

Find out about the ancient Inca and their way of life. Share the information you find with your classmates, and then use what you've learned to write and illustrate a report.

Incan Note Taker

What type of house did they build/live in?	What did they eat?	What did they wear?

What do we remember them for?

•

•

•

•

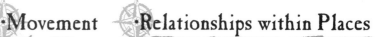
Indigenous Cultures

There are many indigenous cultures in South America. Many contemporary people trace their roots back to the Incas or other indigenous people. Some groups are trying to combine elements of traditional and modern lifestyles. There are smaller groups of indigenous peoples still living a traditional lifestyle.

Select one of these groups of people to learn more about (or select another group indigenous to South America). Take notes about what you learn to share with the class.

Yanomamo	Quechua	Waorani	U'wa
Aymará	Jívaro	Campa	Shipibo

Indigenous Cultures Note Taker

Name of group

Where do they live?	How do they dress?	What type of house do they build/live in?
		What types of foods do they eat?

What interesting customs do they have?

-
-
-
-

The Languages of South America

People living in the different countries speak different languages. The inhabitants of one country may speak many different languages. However, each country also has an official language. Complete this chart by filling in the official language and some other languages spoken in each country.

Country	Official Language	Other Languages Spoken
1. Colombia		
2. Ecuador		
3. Peru		
4. Venezuela		
5. Suriname		
6. Guyana		
7. French Guiana		
8. Bolivia		
9. Paraguay		
10. Brazil		
11. Chile		
12. Uruguay		
13. Argentina		
14. Falkland Islands		

15. What is the official language of most countries? _____

16. Which countries have more than one official language? _____

_____

17. What language of the Incas is still spoken by many South Americans?

_____

Bonus

What is the official language of the country in which you live? Do you speak any other language(s)?

Celebrate Learning

Choose one or all of the following activities to celebrate the culmination of your unit on South America. Use the activities to help assess student learning.

Have a Portfolio Party

Invite parents and other interested people to a "portfolio party" where students will share their completed portfolios, as well as other projects about South America.

Write a Book

A student can make a book about South America. It might be one of the following:
* an alphabet book of South American people, places, or plants and animals
* a dictionary of words pertaining to South America
* a pop-up book of the unique animals of South America

Interview a South American

A student can interview someone from South America or someone who has visited there. The interview could be presented live, as a written report, or videotaped to share with the class.

Create a Skit

One or more students can write and present a skit about an interesting event or period in South American history.

Paint a Mural

One or more students can paint a mural showing one region of South America. A chart of facts about the region should accompany the mural.

Share an Artifact Collection

Students can bring in one or more artifacts representative of South America such as a piece of clothing, a musical instrument, a toy llama, a type of food from South America, etc. A written description of each artifact should be included in the display.

Name _____  Summary of Facts

South America

Relative location _____

Number of countries_____

Continent land area _____

Largest country by area

_____

Smallest country by area

_____

Continent population _____

Largest country by population

_____

Smallest country by population

_____

Highest point _____

Lowest point _____

Longest river _____

Largest island _____

Interesting facts about the continent's regions:

• _____

• _____

• _____

• _____

• _____

Interesting facts about the people:

• _____

• _____

• _____

• _____

• _____

Interesting facts about the plant and animal life:

• _____

• _____

• _____

• _____

• _____

Name

What's Inside This Portfolio?

Date	What It Is	Why I Put It In

Name

My Bibliography

Date	Title	Author/Publisher	Kind of Resource

Search

Name the longest river in South America.

How long is it?

1

Search

The Andes Mountains are found in which South American countries?

2

Search

Name the two countries in South America that are totally surrounded by land.

3

Search

The world's highest waterfall is in South America.

Name the waterfall and tell how high it is.

4

Search

The driest place on Earth is in South America.

Name it.

5

Search

Where can penguins be found in South America?

6

Search

Who owns the Falkland Islands?

7

Search

What is the lowest place in South America?

What is its elevation?

8

Search

Which South American countries are completely north of the equator?

9

What is the largest city in South America?

In which country is it located?

10

Name and give the elevation of the highest mountain in the Andes.

11

Giant turtles are found on what South American islands?

12

Name the highest navigable lake in the world.

Where is it located?

How high is it?

How large is it?

13

Name the South American countries that share a border with Brazil.

14

Name the long, narrow country on the west side of South America.

15

Name the South American country that has the largest land area.

What is its size?

16

Which South American countries border the Pacific Ocean?

17

Which South American countries border the Atlantic Ocean?

18

Give the relative location of the Falkland Islands.

19

What is the official name of…
Suriname?
Uruguay?
Brazil?

20

Which country in South America has Portuguese as its official language?

21

Name this animal.

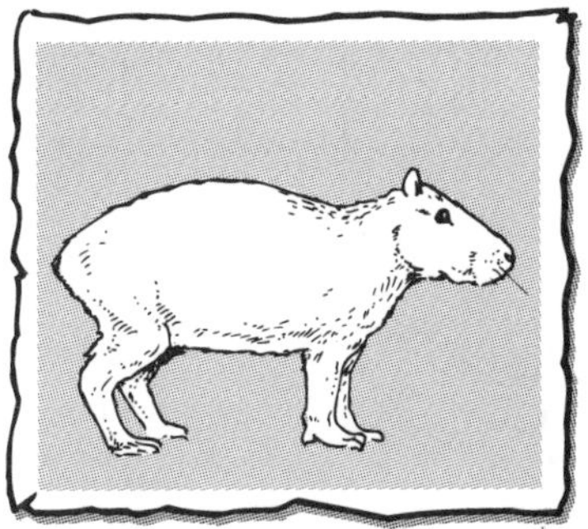

22

What is Machu Picchu?

23

Which continent is closest to the southern tip of South America?

24

Name the continents that are larger than South America.

25

Name the continents that are smaller than South America.

26

What is this plant?

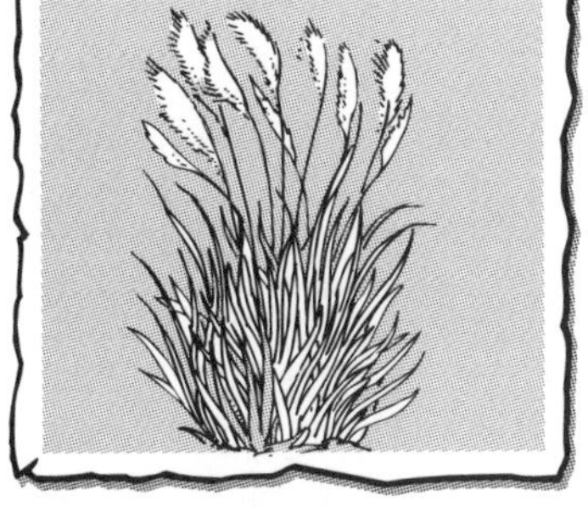

Where is it from?

27

South America

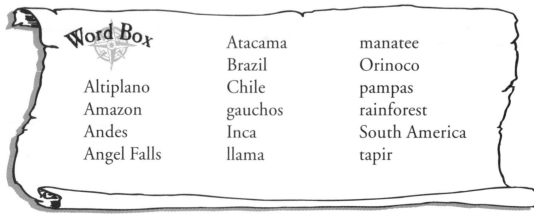

Word Box

Altiplano
Amazon
Andes
Angel Falls

Atacama
Brazil
Chile
gauchos
Inca
llama

manatee
Orinoco
pampas
rainforest
South America
tapir

Across

1. the longest mountain range in South America
3. the largest country in South America
5. a fat, hoofed mammal found in tropical South America
6. an Indian people who ruled an empire in Peru before the Spanish conquest
7. the highest waterfall in the world
9. a large water mammal living in the Amazon River
11. a long, narrow country on the western coast of South America
13. the grasslands of southern South America, especially in Argentina
14. the driest place on Earth is the _____ Desert
15. a river in South America

Down

1. the longest river in South America
2. a continent in the southern part of the western hemisphere
4. a wool-haired South American animal related to the camel
8. a large, high plateau in South America
10. cowboys of the South American pampas
12. a hot, wet area covered in trees

South America

```
A R G E N T I N A C O N T I N E N T P
S C I Q U I T O M N Z C O L O M B I A
O B H G L L A M A O D E L I T E C E R
U U I I U W A G Z T W E T M A N A R A
T E N N L A P R O U D A S A S T P R G
H N T W C E N I N X U R U G U A Y A U
A O O A T A C A M A D E S E R T B D A
M S U R I N G B O L I V I A I D A E Y
E A N G E L F A L L S U N R N R R L E
R I R P E R U O R I N O C O A Y A F C
I R H B R A Z I L L A P A Z M U D U U
C E E G R A N R A I N F O R E S T E A
A S A O F R E N C H G U I A N A G G D
I N G U Y A N A V E N E Z U E L A O O
F A L K L A N D I S L A N D S H O T R
```

Find these words:

Amazon	capybara	iguana	Quito
Andes	Chile	Inca	rainforest
Angel Falls	Colombia	llama	rhea
Argentina	continent	La Paz	South America
Atacama Desert	Ecuador	Lima	Suriname
Bolivia	Falkland Islands	Orinoco	Tierra del Fuego
Brazil	French Guiana	Paraguay	Uruguay
Buenos Aires	Guyana	Peru	Venezuela

Bonus

Underline all of the animals named in the word list.

Glossary

absolute location (exact location)–the location of a point that can be expressed exactly, for example, the intersection of a line of longitude and latitude.

Antarctic Circle–an imaginary line circling the globe at 66.5°S latitude.

altitude–the height of a thing above a given reference point; the height of a thing above sea level.

Altiplano–a plateau region in South America located in the Andes of Argentina, Bolivia, and Peru.

Amerindian–a name given collectively to the indigenous peoples of the Americas.

archipelago–a large group or chain of islands.

capital–a city where a state or country's government is located.

cardinal directions–the four points of a compass indicating north, south, east, and west.

climate–the type of weather a region has over a long period of time.

compass rose–the drawing on a map that shows the cardinal directions.

continent–one of the main landmasses on Earth (usually counted as seven—Antarctica, Australia, Africa, North America, South America, Asia, and Europe).

culture–the shared way of life of a people including traditions, beliefs, and language.

epiphyte–a plant that grows above the ground supported by another plant or object that gets its nutrients and water from the rain, air, dust, etc.

equator–an imaginary line that circles the Earth midway between the north and south poles, dividing it into two equal parts.

ethnic group–a group of people sharing the same origin and lifestyle.

gaucho–a cowboy of the South American pampas.

gulf–a portion of an ocean or sea partly enclosed by land.

hemisphere–half of a sphere; one of the halves into which the Earth is divided—western hemisphere, eastern hemisphere, southern hemisphere, or northern hemisphere.

immigrant–a person who has come from one country to live in a new country.

indigenous–native to an area; originating in the region or country where it is found.

landform–the shape, form, or nature of a physical feature on Earth's surface (mountain, mesa, plateau, hill, etc.).

latitude–the position of a point on Earth's surface measured in degrees, north or south from the equator.

longitude–the distance east or west of Greenwich meridian (0° longitude) measured in degrees.

manufacture–to make a useful product from raw materials.

meridian–an imaginary circle running north/south, passing through the poles and any point on the Earth's surface.

North Pole–the northernmost point on Earth; the northern end of the Earth's axis.

plain–a flat or level area of land not significantly higher than surrounding areas and with small differences in elevation.

plateau–an area of land with a relatively level surface considerably raised above adjoining land on at least one side.

population–the total number of people living in a place.

prime meridian (Greenwich meridian)–the longitude line at 0° longitude from which other lines of longitude are measured.

reef–a ridge of rock, sand, or coral lying just below the surface of a sea.

relative location–the location of a point on the Earth's surface in relation to other points.

resource–substances or materials that people value and use; a means of meeting a need for food, shelter, warmth, transportation, etc.

rural–relating to the countryside.

scale–an indication of the ratio between a given distance on the map to the corresponding distance on the Earth's surface.

South Pole–the southernmost point on Earth; the southern end of the Earth's axis.

strait–a narrow passage of water connecting two large bodies of water.

territory–a region or district of land not admitted as a state (or province) but having its own legislature and an appointed governor.

Tropic of Capricorn–an imaginary line around the Earth south of the equator at the 23.5°S parallel of latitude.

urban–relating to cities.

Answer Key

page 16
1. Venezuela, Colombia, Ecuador, Peru, Bolivia, Argentina, Chile
2. At 22,834 feet (6960 m), Aconcagua in Argentina is the highest mountain in the western hemisphere.
3. Tierra del Fuego
4. potato
5. llama
6. alpaca

page 17
1. 4000 miles (6437 km)
2. a. emergent layer–very tall trees reaching up above the canopy
   b. canopy layer–composed of the crowns of trees almost touching
   c. understory–an area of small trees and brush growing under the canopy
   d. forest floor–dark, with few low-growing plants such as ferns, herbs, and grasses
3. Answers will vary.
4. logging, mining, and farming

page 18
Answers will vary, but could include:
Alike: both are grassy plains, cattle are raised, parts of both areas flood during the rainy season
Different: the eastern part of the Pampas is very fertile and crops are raised there, some of the native animals are different, more of the Pampas is populated

page 20
1. The James flamingo is a large pink bird. It lives around the salt lakes of the Altiplano. It eats algae.
2. The Andean condor is a black and white bird with a 12-foot (3.7 m) wing span. It lives on high cliffs. It eats carrion.
3. The vicuña is the smallest South American camel. It has thick long hair that keeps it warm in the cold weather. It eats grasses.

page 27
Independent Countries:
1. Argentina
2. Bolivia
3. Brazil
4. Chile
5. Colombia
6. Ecuador
7. Guyana
8. Paraguay
9. Peru
10. Suriname
11. Uruguay
12. Venezuela
French Dependency: French Guiana
United Kingdom Dependency: Falkland Islands

page 28
1. Bolivia
2. Venezuela
3. Brazil
4. Guyana
5. Colombia
6. Peru
7. Argentina
8. Chile
9. Ecuador
10. French Guiana
11. Paraguay
12. Suriname
13. Uruguay
14. Falkland Islands

page 31
1. Buenos Aires, Argentina
2. La Paz, Bolivia
3. Brasília, Brazil
4. Bogotá, Colombia
5. Santiago, Chile
6. Quito, Ecuador
7. Georgetown, Guyana
8. Asunción, Paraguay
9. Lima, Peru
10. Caracas, Venezuela

page 36
1. Brazil
2. Chile
3. Paraguay
4. Peru and Bolivia
5. Venezuela

page 40
Answers will vary, but could include:
Natural Resources–precious metals, petroleum, iron ore, minerals, natural gas, lumber, hydropower
Crops & Livestock–coffee, wheat, sugarcane, cacao, potatoes, timber, rice, corn, soybeans, cattle, poultry
Manufactured Goods–textiles, shoes, chemicals, motor vehicles, meat processing, sugar, tires, cement, fish processing, wood and wood products

page 42
1. seed
2. tuber (thick, fleshy underground stem of the plant)
3. sweet green pepper
4. habanero
5. cacao
6–8. Answers will vary.

page 50

Answers will vary, but could include:
giant prickly-pear cactus–Galápagos Islands
bromeliads–Cloud forests on eastern side of Andes
scrub trees–Gran Chaco
grasses–Pampas
cacao tree–Amazon River basin

page 51

pampas grass	tree fern	epiphyte
kapok	cacao tree	bromeliad
liana	monstera	banana tree

page 52

1. sloth
2. piranha
3. jaguar
4. boa constrictor
5. rhea
6. giant tortoise
7. Andean condor
8. giant anteater

page 53

1. vicuña
2. llama
3. in the mountains of South America
4. in North America
5. wool cloth that is made into clothing
Pictures: guanaco   alpaca
            llama   vicuña

page 57

1. a large group or chain of islands
2. giant tortoise
3. Galápagos penguin
4. domesticated animals that have gone wild
5. low ground plants and shrubs; cactus; fruit
6. Ecuador
7. Answer will vary, but could include:
    let nonindigenous animals run wild
    clear and plant land and then abandon it
    overfish an area
    interfere with the nesting areas of indigenous animals

page 61

1. a. Portugal
   b. Spain
2. Answers will vary, but could be: Suriname, Brazil, Peru, Ecuador
3. Answers will vary, but could be:
    a. Chile–Mapuche, Quechua
    b. Brazil–Kayapo, Yanomamo
    c. Ecuador–Waorani, Otavalo
    d. Peru–Jívaro, Aymará
4. Spanish
5. Catholic
6. A gaucho would be on the pampas tending to herds of cattle.
7. Answers will vary, but should include some of the following:
    The poor often live in shacks with no running water, indoor plumbing, or electricity. Jobs, if people have them, are low paying. The middle-class and wealthy population live in homes similar to those in any developed country. They have access to better food, clothing, and medical care.

page 64

	Official Language(s)	Other Languages Spoken (will vary)
1.	Spanish	
2.	Spanish	Quechan, other Amerindian languages
3.	Spanish, Quechua	Aymará
4.	Spanish	
5.	Dutch	English, Sranang Tongo, Hindustani
6.	English	Amerindian languages
7.	French	
8.	Spanish, Quechua, Aymará	
9.	Spanish	Guaraní
10.	Portuguese	French, Spanish, English
11.	Spanish	Mapuche
12.	Spanish	
13.	Spanish	English, Italian
14.	English	
15.	Spanish	
16.	Peru and Bolivia	
17.	Quechua	

page 69

1. Amazon River–4000 miles (6437 km) long
2. Venezuela, Colombia, Ecuador, Peru, Bolivia, Argentina, Chile
3. Bolivia, Paraguay
4. Angel Falls in Venezuela–3212 feet (979 m) elevation
5. Atacama Desert
6. Falkland Islands, Western Galápagos Island, Coast of Peru, South Argentina, Chile
7. United Kingdom
8. Valdés Peninsula in Argentina–131 feet (40 m) below sea level
9. French Guiana, Venezuela, Suriname, Guyana

10. São Paulo, Brazil
11. Mt. Aconcagua–22,834 feet (6960 m)
12. Galápagos Islands
13. Lake Titicaca in Peru and Bolivia–12,500 feet (3810 m) elevation; 3200 square miles (8288 sq km) area
14. French Guiana, Suriname, Guyana, Venezuela, Colombia, Peru, Bolivia, Paraguay, Argentina, Uruguay
15. Chile
16. Brazil–3,300,171 square miles (8,547,403 sq km)
17. Chile, Peru, Ecuador, Colombia
18. Argentina, Uruguay, Brazil, French Guiana, Suriname, Guyana, Venezuela

page 71

19. in the Atlantic Ocean off the eastern coast of Argentina
20. Suriname—Republic of Suriname
    Uruguay—Oriental Republic of Uruguay
    Brazil—Federative Republic of Brazil
21. Brazil
22. capybara
23. ruins of an ancient Incan city
24. Antarctica
25. Asia, Africa, North America
26. Europe, Antarctica, Australia
27. pampas grass–the grasslands of Argentina

page 74

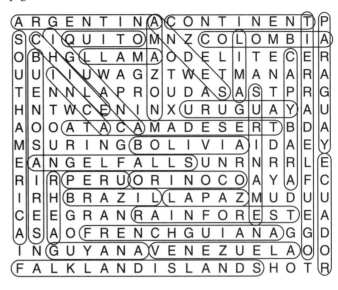

page 72

Bibliography

Books about South America

Amazon Basin by Jan Reynolds; Harcourt Brace Jovanovich, 1993.

Argentina (Major World Nations Series) by Sol Leibowitz; Chelsea House, 1998.

Brazil (Country Fact Files Series) by Marion Morrison; Raintree Steck-Vaughn Publishers, 1994. (Also available in this series: Ecuador, Peru, Bolivia)

Brazil (Countries of the World Series) by Michael Dahl; Bridgestone Books, 1997.

Camels (Zoobooks) by John Bonnett Wexo; Wildlife Education, Ltd., 1989.

Ecuador (Cultures of the World Series) by Erin L. Foley; Marshall Cavendish Corporation, 1996. (Also available in this series: Argentina, Bolivia, Brazil, Chile, Colombia, Peru, and Venezuela)

Peru: The Land by Bobbie Kalman and David Schimpky; Crabtree, 1994.

Postcards From Brazil by Zoë Dawson; Raintree Steck-Vaughn Publishers, 1996.

Rain Forest by Robin Bernard; Scholastic Inc., 1996.

South America by C.J. Carella and Kevin Siembreda; Palladium Books, 1994.

South America (Continents Series) by Ewan McLeish; Raintree Steck-Vaughn Publishers, 1997.

The Grandchildren of the Incas by Matti A. Pitkänen; Carolrhoda Books, Inc., 1991.

The Incas (See Through History Series) by Tim Wood; Viking Childrens Books, 1996.

The Waorani: People of the Ecuadoran Rain Forest by Alexandra Siy; Dillon Press, 1993.

This Place is High: The Andes Mountains of South America by Vickie Cobb and Barbara Lavallee; Walker and Company, 1993.

What Do We Know About the Amazonian Indians? by Anna Lewington; Peter Bedrick Books, 1993.

Yanomami: People of the Amazon by David Schwartz and Victor Englebert; Lotherop, Lee, and Shepard, 1995.

General Reference Books

(Maps and atlases published before 1997 may not have the latest changes in country names and borders, but they will still contain much valuable material.)

Atlas of Continents; Rand McNally & Company, 1996.

National Geographic Concise Atlas of the World; National Geographic Society, 1997.

National Geographic Picture Atlas of Our World; National Geographic Society, 1994.

The New Puffin Children's World Atlas by Jacqueline Tivers and Michael Day; Puffin Books, 1995.

The Reader's Digest Children's Atlas of the World; Consulting Editor: Colin Sale; Joshua Morris Publishing, Inc., 1998.

The World Almanac and Book of Facts 1998; Editorial Director: Robert Famighetti; K-III Reference Corporation, 1997.

Technology

CD-ROM and Disks

Encarta® Encyclopedia; ©Microsoft Corporation (CD-ROM).

MacGlobe & PC Globe; Broderbund (disk).

Where in the World Is Carmen Sandiego?; Broderbund (CD-ROM and disk).

World Fact Book; Bureau of Electronic Publishing Inc. (CD-ROM).

Zip Zap Map; National Geographic (laser disc and disk).

Websites

For sites on the World Wide Web that supplement the material in this resource book, go to http://www.evan-moor.com and look for the Product Updates link on the main page.

Check this site for information on specific countries:
CIA Fact Book–www.odci.gov/cia/publications/factbook/country-frame.html